Storytime Discoveries

Read-Aloud Stories and Demonstrations About
Math

Written by Dotti Enderle
Illustrated by Ginger Illustration

Teaching & Learning Company
1204 Buchanan St., P.O. Box 10
Carthage, IL 62321-0010

This book belongs to

____

ISBN No. 1-57310-440-X

Printing No. 987654321

Teaching & Learning Company
1204 Buchanan St., P.O. Box 10
Carthage, IL 62321-0010

Table of Contents

Dear Teacher or Parent,

It stands to reason that learning is more effective when the teaching tools are fun. That's why I developed *Storytime Discoveries: Math*. This collection introduces students to various mathematical concepts, while entertaining them with interactive stories and activities.

Storytime Discoveries: Math is a collection of original stories, folktales and poems, designed to encourage logical thinking and problem solving from a child's perspective. The stories are well rounded, giving them an edge over regular word problems. Each comes with simple instructions, most needing no extra materials. Whether planned as a lesson or used as a filler, they provide added diversity to the math curriculum.

Let's face it, math is fun. And this collection will help you convey that to your students. Math and storytelling. It all adds up.

Sincerely,

Dotti Enderle

The Land of Equal

two glass jars
box of salt
coffee scoop
small bowl

Measurement

Pour and scoop salt as instructed within the story.

The Land of Equal

Many years ago in the Land of Equal, EVERYTHING was equal. The number of men equaled the number of women. *(Pour an equal scoop of salt into each jar.)* The number of horses equaled the number of donkeys. *(Pour another scoop of salt into each jar.)* The number of dogs equaled the number of cats. *(Repeat.)* And the number of birds equaled the number of fish. *(Repeat.)* Everything was equal in the Land of Equal, and the people were equally happy, especially their ruler, King Steven of Even, who always kept a level head.

But one day thunder roared in the distance. A cloud of smoke covered the sun.

"Look!" someone shouted.

A shadow fell over the kingdom as the evil dragon, Lop Side, approached.

The people scattered and ran for cover. The squirrels hid in the trees. The snakes hid in the grass. And the groundhogs stayed buried in their holes. Everyone and everything was equally afraid of Lop Side.

Lop Side stood over the Land of Equal like a great mountain. He sniffed the air and let out a sigh. Smoke curled from his nose. Then Lop Side scooped up one of the cats. *(Scoop salt from one of the jars and pour it into the bowl.)*

"No!" King Steven of Even cried out as he watched from a castle window.

Lop Side scooped up one of the dogs. *(Scoop more salt from the same jar and pour into the bowl.)* And just before he turned to go, Lop Side reached in through the castle chimney and scooped up the king's daughter, Princess Pitter-Patter! *(Repeat.)*

King Steven of Even pounded his fists and kicked his feet. "No! No! No!"

Now the Land of Equal was very UNEQUAL.

King Steven of Even made a proclamation. He would share his gold equally with anyone who could capture the dreaded dragon, Lop Side, and return the princess and her pets back to the kingdom.

The people read the proclamation and backed away, shaking their heads. No one was brave enough, tough enough and smart enough to equal the monstrous dragon—except for one man.

"I'll rescue them!" shouted a young knight named Sir Fifty-Fifty. He jumped on his horse, drew his sword and charged off. Sir Fifty-Fifty rode for two days, splashing through streams, whacking through thickets and tottering near the edge of steep cliffs. Finally, he arrived at the dark cave of the dragon, Lop Side.

He quietly sneaked in and took a peek. The cat cowered in the corner. The dog shivered in the shadows. Princess Pitter-Patter wept a river of tears. And Lop Side laid sleeping, smoke curling from his nose.

"Shhhh," Sir Fifty-Fifty said, pressing his finger to his lips. He motioned for them to come.

The dog and cat tiptoed toward him, but Princess Pitter-Patter kept sobbing. She never looked up.

Sir Fifty-Fifty snapped his fingers, then waved his arms. Princess Pitter-Patter continued to cry.

"Hey, do you want to be rescued or not?" he whispered to her.

She jerked her head up and saw Sir Fifty-Fifty inching toward her. Princess Pitter-Patter ran straight into his arms.

Just as they were about to sneak out, a large claw dropped down before them, blocking their way. They turned to see Lop Side looming over them.

He let out a great roar.

Sir Fifty-Fifty quickly slipped off one of his boots, and when Lop Side lunged at him, Sir Fifty-Fifty used the boot to bop him right in the nose!

Lop Side hopped around, moaning and groaning, patting his swollen nose.

He turned back to Sir Fifty-Fifty. His eyes glowed like rubies. But Sir Fifty-Fifty used his boot to scoop up the mud that had been made by Princess Pitter-Patter's tears.

And just as Lop Side was about to let out a roar that would turn him to toast, Sir Fifty-Fifty poured the mud straight down Lop Side's fiery throat. Lop Side coughed and sputtered, and when he tried to roar, he only made globs of messy mud pies.

"I guess we're equal now," Sir Fifty-Fifty said as Lop Side slunk into the corner to hide.

So Sir Fifty-Fifty brought back the cat, *(Scoop salt from the bowl and pour it back into the jar.)* and the dog, *(Repeat.)* and the princess. *(Repeat.)* And the Land of Equal was equal once again.

King Steven of Even shared his gold equally with Sir Fifty-Fifty, and to celebrate, the kingdom held a royal picnic.

Everyone was equally happy. Until King Steven of Even accidentally swatted a mosquito. *(Scoop some salt out of the jar.)*

"OOPS!"

Shapes for Sale

chalkboard
chalk

Shapes

Draw the shapes on the chalkboard as you read the story.

Shapes for Sale

Melissa loved Christmas. She wanted this one to be special. She decorated her Christmas tree with red balls, gold stars and silver garland. But Melissa wanted something different for the top. So, she went into town to do some shopping.

Melissa passed store after store, each dressed up with sparkling displays and shiny tinsel. But as beautiful as everything was, Melissa still couldn't find what she wanted. Then she came to a plain brown building with a sign in front that read: SHAPES FOR SALE. Melissa hurried in. Next to the checkout counter she found a large bin filled with every shape you can imagine. Melissa dug in.

The first shape she found was a triangle.
(Draw a triangle on the board.)

She held it up. "This is great," she said. "I love the way all three sides are the same size."

Next, she found a figure eight. "Is there no end to this?" she asked as she traced her finger in and out and around it. She set the figure eight right down on top of the triangle. *(Draw a figure eight sideways above the triangle.)* It balanced nicely.

Then, Melissa found a circle. "Oh my!" she cried. "This is the roundest circle I've ever seen!" Melissa rolled the circle around on the floor, then picked it up and placed it on top of the figure eight.
(Draw a circle above the figure eight.)

Melissa continued rummaging through the bin. She found something she thought was an egg. After a closer look, she saw that it was an oval. "Ovals are such fun shapes," she said. "I really love this one." She put the oval right above the circle.
(Draw the oval just above the circle to look like a halo.)

Melissa stepped back. A look of surprise crossed her face. "Well, aren't you an angel!" she said. Melissa paid for her angel and hurried home. She placed it right on the very top of her Christmas tree.

"This is shaping up to be the best Christmas ever!" she said.

When the Numbers Go Marching In

To the tune of "When the Saints Go Marching In"

Number recognition

Oh when the ones go marching in.
Oh when the ones go marching in
I want to add and subtract that number.
When the ones go marching in.

Oh when the twos go marching in.
Oh when the twos go marching in
I want to add and subtract that number.
When the twos go marching in.

Oh when the threes go marching in.
Oh when the threes go marching in
I want to add and subtract that number.
When the threes go marching in.

Oh when the fours . . . fives . . . sixes . . . sevens . . .
eights . . . nines . . . tens . . . .

Cinderella

A Time Tale

reproducible page 13
construction paper
brad

Telling time

Photocopy the picture of the clock. Cut clock hands from construction paper and fasten with a brad. Each time Cinderella looks at the clock, ask the students where the hands should be on the clock and move them to the appropriate spots.

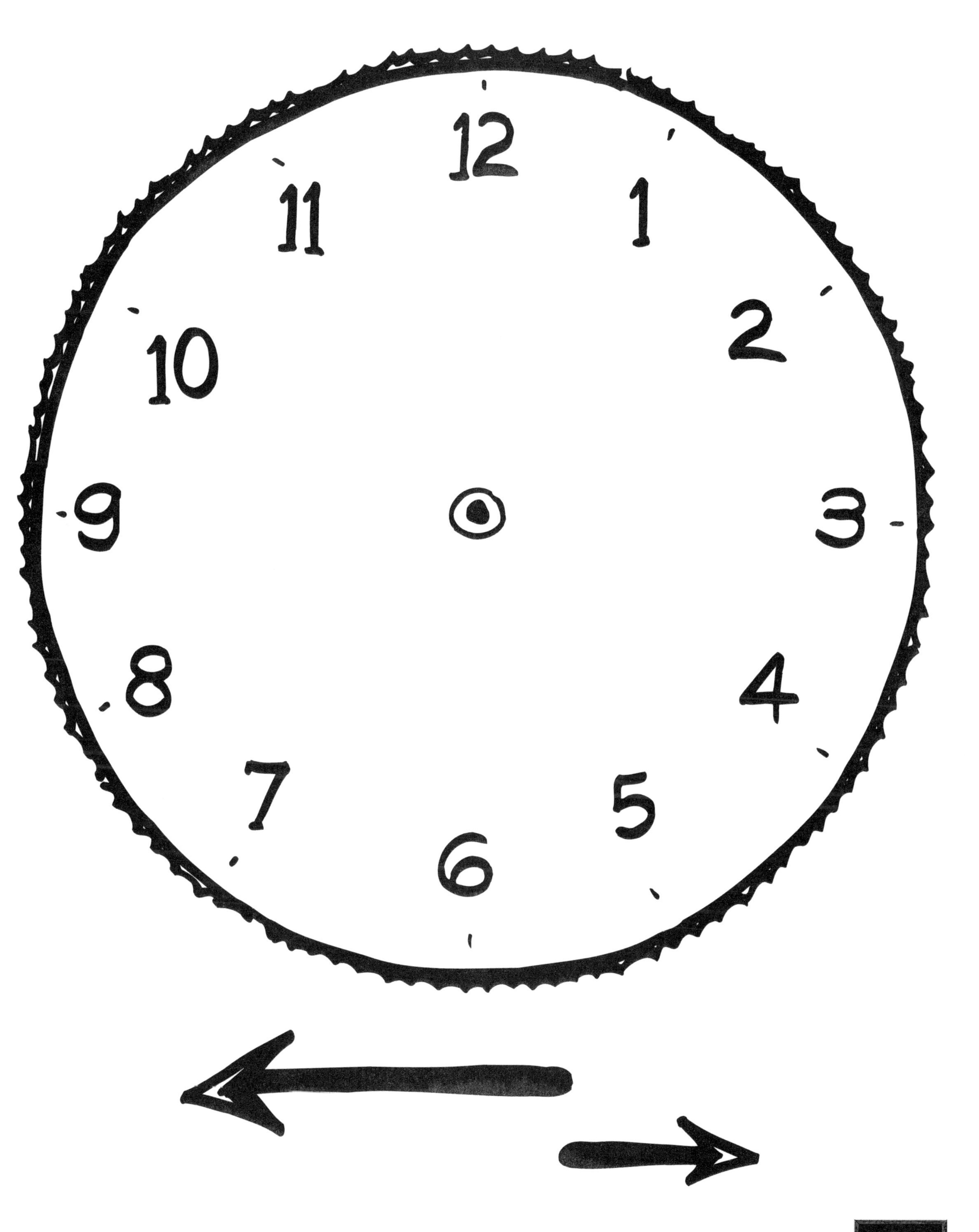
12
1
2
3
4
5
6
7
8
9
10
11

Cinderella

A Time Tale

Cinderella adjusted the hands on her clock. Five o'clock exactly. Two hours until the royal ball, and she still had work to do. She cooked and cleaned the kitchen. Swept up the crumbs, then went to help her ugly stepsisters get dressed. As she climbed the stairs, she looked back at the clock. Six o'clock. All those chores had taken a full hour.

Her stepsisters waddled and prattled and took their sweet time about getting ready. Cinderella watched the minutes tick away as she tied sashes, fastened beads and clipped in fancy hair combs. At last, the stepsisters were ready. Well, as ready as those ugly stepsisters could be. If she'd had all the time in the world, Cinderella couldn't make those two look pretty.

"Come girls," her stepmother said. "The carriage is waiting."

"But I haven't had time to get ready!" Cinderella said.

Her stepsisters burst into ridiculous fits of laughter, snorting and sniffling.

"You only need time to get our bed warmers ready for when we return," Stepmother said. "Although I'm sure we'll be too excited to sleep. I just know the prince will pick one of these two lovelies for his bride."

Only if he's blind, Cinderella thought.

Her stepmother and stepsisters brushed by her as they hurried out and loaded themselves into the waiting carriage. Cinderella sulked as she trudged down the stairs. She passed by the clock while walking through the kitchen on her way out to the garden. Seven o'clock. The trumpets would be sounding. The grand hall filling up with ladies in beautiful gowns, all hoping to dance with the prince. But there would be no dancing for Cinderella. She plopped down on a bench outside and lowered her head, sobbing.

"Stop that blubbering!" a voice called out.

Startled, Cinderella hopped up. A small plump woman stood beside her. "Who are you?"

"I'm your fairy godmother," the lady said. "And it's about time I showed up, too." She looked at her watch. "It's already 7:15. You'll be late for the ball!"

"I can't go to the ball! Even if I had time to get ready, I have no way to get there."

The fairy godmother grunted. "Then we'll have to do this lickety-split." She waved her magic wand over Cinderella, and with a breath of wind, her dress turned into a pink satin ball gown. Her hair was tied up in flowers and ribbons, and her tiny feet were pressed into a pair of crystal glass slippers. On her wrist was a dainty jeweled watch.

Another quick wave of the wand and a pumpkin in the garden turned into a carriage. A couple of playful mice became the horses.

"Nothing to it," the fairy godmother said. "You'll be at the ball in no time at all."

"This is wonderful!" Cinderella said. "I may dance all night!"

The fairy godmother took Cinderella by the hand. "Not all night, dear. You only have until midnight, then the spell runs out. Keep an eye on your watch. Just make sure you're home by then."

Cinderella smiled. She didn't mind the curfew. She was just thankful to be going.

"My goodness, look at the time!" the fairy godmother exclaimed. "7:20! You must be off!"

Cinderella stepped into the carriage, and waved good-bye as the horses pulled away, swiftly taking her to the palace.

The clock tower chimed just once. Cinderella looked down at her watch. 7:30. The palace was lit like a wonderland, enchanting and beautiful. She stepped out of her carriage and entered through the towering entrance. All eyes turned. Time seemed to stand still. And the prince smiled and stepped forward. "May I have this dance?"

Her heart beat with every tick of the clock as she floated across the ballroom, the prince twirling her round and round. But she was careful. After a while, she looked at her watch. 8:00. Great! She didn't have to be home for four hours.

The night was like a fairy tale. The prince refused to dance with anyone but Cinderella. As they swept past, her ugly stepsisters scoffed and turned up their noses. Her stepmother sneered. Cinderella could practically hear her growling under her breath. But she was sure they had no idea who she was. In all the time she'd lived with them, they'd never seen her look this radiant.

The evening flowed like the music in the ballroom. Cinderella did remember to peek at her watch once more. 9:15. Still plenty of time. But shortly after that is when it happened. Cinderella and the prince stepped out onto the balcony, just the two of them, and he gave her a tender kiss. Time ceased to exist for her. She didn't want to think about midnight . . . or tomorrow . . . or forever. She only wanted this night to never end.

But it did. She forgot to look at her watch, and before she knew it, the clock struck 12 times. Panicked, Cinderella fled through the palace, rushed out to the carriage and hopped in. The horses only took her a short distance when the clock stopped striking. She found herself miles from home, wearing her raggedy dress and sitting next to a pumpkin and two mice. She hung her head and cried.

After walking the long way home, she sneaked into the back door. She knew she'd be in trouble for not having her stepmother and stepsisters' beds warm for them when they'd returned. But there would be time for excuses in the morning. She fell across her bed, half sad, half breathless. It had been the most beautiful night of her life. She glanced at the clock by her bed. 1:00 a.m.

She woke to an uproar of bickering and hollering. "Cinderella! Where's our breakfast! Do you have any idea what time it is?" She didn't. Looking at the clock, she could see it was already 8:00. She hurried downstairs to the kitchen.

"Did you see her?" one stepsister asked the other. "That dress she wore made her look like a wilted rose!"

"And what about those glass shoes she wore?" the other chimed in. "I bet they weren't real crystal."

"Oh hush!" Stepmother shouted as she sat down. "The silly girl disappeared, didn't she? I heard that the prince never even found out her name. He can search till the end of time, but he'll never find her. Then he can pick one of you to wed."

Stepmother was right. Cinderella was so enchanted by the prince, she never took the time to tell him her name. He'd never know who she was.

She hadn't realized that when she rushed off in a frenzy, she'd dropped one of her glass slippers on the stairs. The prince had picked it up, and vowed to marry the girl who's foot could fit into the teeny shoe.

The day passed slowly . . . 10:00 . . . 11:00 . . . noon. No matter how busily Cinderella worked, time still crept like a snail. Her stepsisters were in a particularly foul mood, and made her day even more miserable. But still the afternoon came and went. 1:00 . . . 1:30 . . . 2:00 . . . 2:30 . . . 3:00 . . . 3:15 . . . 4:00 . . . on and on. And just about dinnertime, when the sun was setting in the west, they heard a commotion outside. Horses trotting. Boots stomping. Then a knock at the door.

"It's the prince!" Stepmother called. She shuffled her daughters toward the parlor. "Not you!" she told Cinderella. "Keep yourself hidden here in the kitchen."

Cinderella opened the kitchen door just a crack, and peeped through. There he was, looking magnificent. But his face seemed worried and frantic.

"Ladies," he said with a bow. "I'm searching for the owner of this shoe. Would each of you please try it on?"

The ugly stepsisters nearly plowed each other down trying to get to it.

"Girls," Stepmother said. "One at a time. She sat the oldest (and ugliest) down first. The shoe wouldn't even fit on her big ogre toe.

The other stepsister tried it on. She crammed so hard, her foot and her face turned a sickening shade of purple.

Cinderella giggled to herself when her stepmother tried it on. The prince cringed as she took her time, squeezing only her toes inside.

"Thank you for your time, ladies," the prince said. "I'll be going."

Cinderella knew it was now or never. She pushed open the kitchen door. "What about me?"

The prince knelt down. Then taking Cinderella's tiny foot in his hand, he slid the slipper on, as soft as silk. Stepmother fainted.

The clock struck 12:00. Straight up noon. The orchestra played. The crowd rose. And Cinderella walked down the aisle of the church, wearing both glass slippers (a gift from you-know-who). In no time at all, they were married. And of course, they lived happily ever after.

Name ______________________________

Trading Coins

Cut out the coins and glue them into the correct boxes.

1. If I have five pennies, and I'm not feeling fickle,

   I trade them all for a shiny round  .

2. If I have two nickels and plenty of time,

   I can trade them for a thin silver  .

3. Two dimes and a nickel (not just in that order)

   Can always be traded for a convenient  .

4. You now have four quarters? Get ready to holler.

   Trade them all in for a crisp paper  .

The Shortest Route

reproducible page 21

Measuring

Make enough copies of the sign for each child. Let them decide which line is shortest. Then have them use a ruler or length of string to measure both lines.

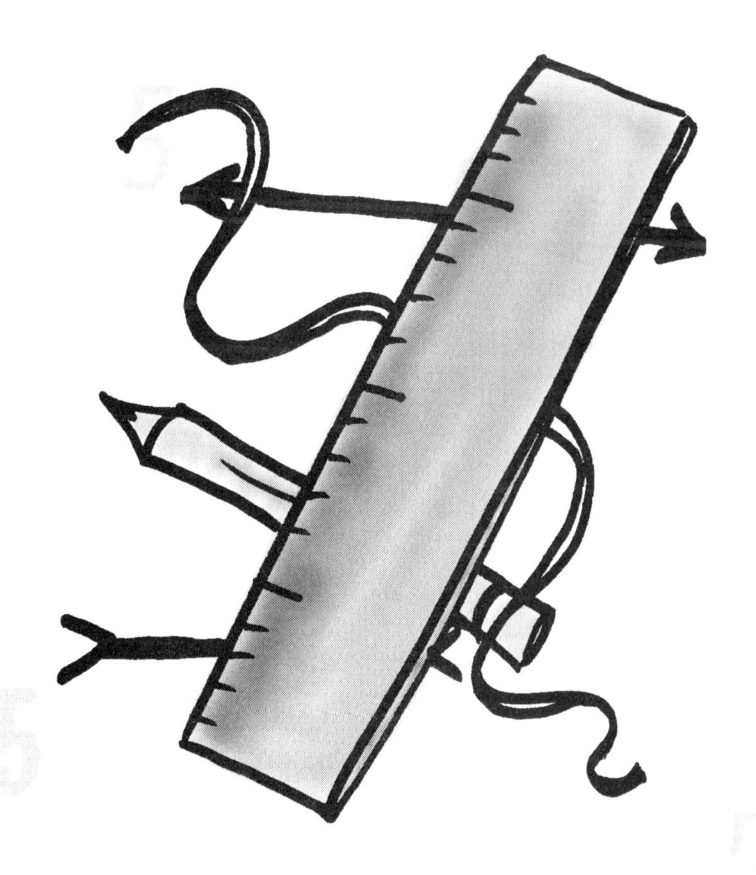

The Shortest Route

Rudy Roo lived in the town of Oodles. Oodles was a strange place with upside down trees, backward cars and houses built on top of bridges.

One day Rudy decided it was time for a vacation. He packed his bag, hopped into his car and skedoodled out of Oodles. He headed west because he longed to see the town of East Grubbly, known for its southern cooking. He toddled along, thinking of how much fun he'd have.

He hadn't gone far when he came to a crossroad with a sign.

EAST GRUBBLY

Right <——————————————>
Left >——————————————<

Hmmm . . . naturally, Rudy wanted to go the shortest distance to get there. He studied the sign carefully. Left or right? It seemed obvious which direction to turn. So he made his decision and rambled on to East Grubbly, getting there after dark.

Rudy turned on the headlights of his cars. Bummer! he thought. I went the shortest way, but it still took so long to get here. It is already dark out.

Rudy stayed a few days in East Grubbly, and enjoyed some fine meals. But when it was time to go home, he looked at the sign at the edge of town.

OODLES

Right >——————————————<
Left <——————————————>

Rudy took the shortest route home. Or did he? He still got there after dark.

Rudy decided to stay in Oodles for a long time without any more special trips out of town. He was just no good at judging distance.

Name ___________________________________

The Shortest Route

Flapjack Stack

blocks
books
anything that's
stackable

Addition and logic

As Benny makes flapjacks, stack the objects on top of each other to represent the flapjacks. Have the kids guess how many flapjacks Benny can eat on an empty stomach.

Flapjack Stack

"I'm so hungry," Benny said as he sat up in bed and stretched. Benny's tummy rumbled like a cloud of thunder. "Oh, my stomach is so empty."

Benny stumbled out of bed and went into the kitchen. He heated the griddle, stirred up some flapjack mix and thought, "How many flapjacks can I eat on an empty stomach?"

His tummy said, "Grrrrrrrr."

Benny poured out three round thick flapjacks. He stood by as they bubbled. Using the spatula, he flipped the flapjacks in the air, and they landed back on the griddle, golden side up. "Mmmm . . . that smells good," he said.

His tummy said, "Grrrrrrrrr."

Benny swept the flapjacks up with the spatula and stacked them on a plate. Somehow it didn't look like enough. "I wonder how many flapjacks I can eat on an empty stomach?"

His tummy said, "Grrrrrrrr."

Benny poured four more flapjacks on the griddle. This time, he flipped them even higher in the air, feeling confident, though hungry. When they were done, he put them on the stack. "Now how many are here?" Benny counted them. "They look delicious," he said, "but they are small. I wonder how many I can eat on an empty stomach?"

His tummy said, "Grrrrrrrr."

Benny wanted to make sure he filled his tummy up. So he poured five more flapjacks on the griddle. When it was time to turn them over, he flipped them so high, they nearly hit the ceiling. But they landed—plop—right back on the griddle.

When they were done, Benny stacked them on top of the others. "How many do I have now?" He counted them. "I guess I'm about to find out how many flapjacks I can eat on an empty stomach."

His tummy said, "Grrrrrrrr."

He buttered his flapjacks, added some fresh strawberries and maple syrup and dug in. "Mmmm . . ." he said.

His tummy didn't say a thing.

So . . . how many flapjacks could Benny eat on an empty stomach?

Answer: Just one. After eating one, his stomach was no longer empty.

St. Ives

A Mother Goose Rhyme

As I was going to St. Ives,
I met a man with seven wives.
Every wife had seven sacks;
Every sack had seven cats:
Every cat had seven kits:
Kits, cats, sacks, wives,
How many were going to St. Ives?

Six Slices or Eight?

reproducible page 28
pencils
rulers

Fractions

Give each child a photocopy of the pizza page. Using the pencil and ruler, have them divide the pizza in half, then four slices, then eight.

Six Slices or Eight?

When John walked into the pizza parlor, the smell of warm rising pizza filled his nose and made his mouth water. A wonderful cheese and pepperoni pizza was just waiting for him and his three friends, Eric, Charlie and Nick.

John watched as the man lifted it from the oven with a large pizza shovel. Yum!

The man transferred the pizza into a cardboard box and asked, "Six slices or eight?"

John thought about it. He only needed four slices. He almost said so, then thought about how big that one slice would be. He'd have to lift it with both hands, and that would still be awkward. He imagined the small circles of pepperoni rolling off onto his plate.

"Young man," the pizza man said. "Six slices or eight?"

John scratched his head. Maybe six? But that wouldn't work. If they each took a slice, then there would be two slices left over. After all, six minus four equals two. Of course they could divide the two remaining slices in half. Two slices, cut in half, would equal four more.

"Uh-hum," the man said, impatiently tapping the pizza cutter on the counter. "Six slices or eight?

John wished his friends were here to help. But then he thought about the pizza cut into eight slices. They could each take one slice, and there would be four slices left. Eight minus four equals four. And that would mean they'd each get another full slice!

"Six or eight?" the man repeated once more.

"Eight!" John said. "Two slices for me and two for each of my friends."

The man cut the pizza into eight slices and closed the lid on the pizza box.

John paid for the pizza with the money he'd collected from his friends. He ran out to meet them at the picnic table in the park.

"Wow, this looks great!" Eric said.

"I can't wait to dig in!" Charlie shouted.

"Delicious," Nick mumbled through a mouthful of pizza.

They each ate their two slices, washing it down with a thermos of lemonade.

"Now," Charlie asked John, "can you divide up our change?"

Name ________________________________

Six Slices or Eight?

Of all the food, nothing beats—a,
chewy slice of cheesy pizza!

Little Red Riding Hood

reproducible page 32

Map reading

Follow the directions for "Little Red's Map."

Little Red Riding Hood

There once was a girl called Little Red Riding Hood. Why? Because her granny had made her a lovely red cape with a hood, and Little Red wore it every time she went out. When everyone else wore brown or black or gray, Little Red stood out in her red riding hood.

But there came a time when her granny became ill and weak and had to stay in bed for a several days. Little Red's mother packed a basket of goodies and sent her through the woods to deliver them to Granny.

"Here's a map," Mother said. "Use it to find your way, and you won't have any problems getting there."

Little Red looked at the map. She noted the best route to Granny's and headed off. She walked north on Pine Trail for a half mile and turned west onto Rainbow Path. She loved Rainbow Path, named for all the colorful flowers that grew there. She only walked a short distance, but could no longer resist the beautiful bouquets just off the road. She knew she should continue on as the map showed. "But wouldn't Granny love some nice flowers to brighten up her day?" Little Red said to herself. She wondered into the field and began picking.

A scruffy old wolf peeped out from behind Enchanted Rock. "Why are you wandering out here all alone?" he asked Little Red Riding Hood.

She smiled up at him. "My granny is sick. I'm taking her some food in this basket."

The wolf looked around. "Your granny lives way out here?"

"Oh no, no," she said, laughing. "Granny lives on Sparrow Lane. See, here's the map." She drew an imaginary line with her finger, showing the route she would take.

The wolf looked down at it. Little Red was going the long way.

Hmmm . . . he thought. I could gobble this little girl up right now while no one's here, but she's so small and scrawny. I think I'll just meet her at Granny's house. That way I can have lunch and dessert.

That sly old wolf took the shortest route to Granny's, and got there long before Little Red. He crept in, tied Granny up and put her in the closet. Then dressing up in her nightgown and kerchief, he slipped into her bed and covered up.

Meanwhile, Little Red Riding Hood checked her map. She continued west on Rainbow Path, turned north at the Weeping Willow, and followed that trail for a quarter mile to Happy Hollow. She turned east. It was just a hop, skip and a jump to Granny's house.

Little Red Riding Hood knocked on Granny's door. "Granny, it's me, Little Red."

"Come in, child," the sneaky old wolf said in a high-pitched voice.

Little Red put the basket on the table and the flowers in a vase of water. She walked over to the bed and looked down at Granny, smiling up at her with a slobbery grin. "Granny, what big eyes you have."

"The better to see you with, my dear," the wolf said. Although he thought, the better to see a shortcut on a map.

"Granny, what big ears you have."

The wolf wiggled them, nearly knocking the kerchief off his head. "The better to hear you with, my dear."

Little Red Riding Hood took a step back and stuttered, "Granny, what big t-t-teeth you have!"

"The better to eat you with!" The wolf jumped out of bed and chased Little Red Riding Hood around the house. She didn't need a map to find her way now. She ran into the next room, hid behind the door and waited.

"Little Red Riding Hood?" the wolf sang out. "Where are you?" He slipped in through the door, but before he could duck, Little Red whacked him over the head with a table lamp, knocking him to the floor.

"Take that!" she said. The wolf was out cold.

She rescued Granny from the closet, and the two of them dragged the wolf out and rolled him off the front porch. "And stay out!" Little Red Riding Hood added. She locked the door tightly, put Granny back to bed and the two of them enjoyed the basket of goodies she'd brought.

Name ______________________________

Little Red's Map

Using a green crayon, color the route Little Red Riding Hood took.
Color the wolf's route with a red crayon.

Greater Than Jack

chalk
chalkboard

Greater than, less than

Draw the symbols on the chalkboard.

Greater Than Jack

I learned something new at school, and it's great for a secret code. Math symbols. > means "greater than." So 10 > 9. And < is "less than." So 8 < 9. Now I have a new way of getting back at Jack!

Jack thinks he's so much better than me, but he's not. I can outrun him in relays. I can out sing him in music. And I can definitely stand on my head a lot longer. Jack just tumbles over like a piece of limp spaghetti. Yet he still runs around saying, "I'm better than Jenna!" Poo! If he's better than me, then why doesn't he know the secret code?

Jenna > Jack.

I write it everywhere. On my notebooks, the chalkboard, in the dirt on the playground. I want everyone to know. But Jack just doesn't get it. Yesterday at lunch he came up to me, acting tough and mad. "Stop writing that!"

"Why?" I asked. "You don't even know what it means."

"Yes, I do. It means 'love.' Jenna loves Jack."

"Ewwwwwwww! Are you kidding? Gross me out. It means 'I'm greater than you.'"

Jack's face turned the most mysterious shade of purple I'd ever seen. "No way! I'm greater than you!"

So for the rest of the day Jack wrote Jack > Jenna everywhere. And, of course, he took the time to erase what I'd written . . . except on my notebooks.

"Jack, you are not greater than me, and I can prove it."

"Okay," Jack said. "Prove it."

"Fine. We'll see who's the best at math."

Jack rolled his eyes. "No sweat. You're a girl. Girl's don't know math."

That did it. I wasn't just a girl. I was me. Me! My ponytail didn't pull my brains out. "Oh yeah! I'll show you."

Lucky for us, the next day, Mrs. Smith, our teacher gave a math test. I glared at Jack. Jack glared at me. You could cut the tension with a pair of scissors. I just wish I hadn't told Jack what > meant. Now he would know for the test.

I had learned a trick that the greater than, less than sign looked like an alligator. And it always bit at the greater number. I was betting Jack didn't know that.

The test was tough. There were lots of three-digit addition problems. Some subtraction. Some tricky word problems. And a killer bonus problem. Ugh. This was a true test!

I began to sweat. I held my concentration, and double-checked my work . . . triple-checked it! I'm just glad it wasn't a timed test. I figured I'd be the last to turn in my exam. But I wasn't. Jack finished at the exact same time as me. Bummer. He'd triple-checked, too. Or maybe not. Maybe he was just so dumb it took him forever to work the problems out the first time. I was counting on it.

It would be an eternity before tomorrow came and I found out who got the better test score. I walked home after school seeing my secret code written everywhere. Only Jack had written it.

Jack > Jenna.

Jenna < Jack.

I kicked the dirt and tried not to let it upset me. The rest of the afternoon and evening was a wash out. All I could think of was the test! Was the answer to number 12 really 16? Did Billy travel 200 miles to Grandma's in problem 24? Is Jenna really > Jack?

By the time I got to school the next day, I was a wreck. Jack was cool as a penguin, sitting with a smirk on his face. Had he seen the test scores?

Mrs. Smith went through the usual routine of the day. English first—we were working on verbs. Then science—how long does a butterfly stay in the chrysalis? Then social studies. I tried to do my best, but it was like having ants in my pants. I wiggled all morning waiting for math.

Lunch and recess calmed me down some, but finally we lined up for the bathrooms and water fountain, and I knew after 20 minutes of silent reading, we'd be heading straight for math. Jack bumped against me from the boys' line and mouthed "I'm greater than you."

Boloney!

The moment of truth came. My body was as tight as a wound rubber band. I could see a few sweat beads on Jack's forehead, but wasn't sure if it was from nerves or playing kickball outside. The air in the room was thick. Then it happened . . .

"I have your test grades here," Mrs. Smith said, waving them toward us. She passed them out. I waited, drumming my fingers on my desk. She took her time, licking her thumb before sorting them, just so two wouldn't stick together. She strolled up and down each aisle. Everyone looked at their scores, some smiling, some wrinkling their noses. She was down to the last two papers.

"I may be an old fuddy-duddy," she said, "but I'm not blind. I make it a point to know what's going on in my classroom. I've never cared for competition. People are all different. Some excel at one thing while another at something else. But I have to say, I'm pleased with the results that some competitions can supply."

She didn't say another word. Jack and I looked at each other as she handed us our tests. We each made a 100 plus 5 points for the bonus question.

Then Mrs. Smith went to the chalkboard and wrote

Jenna = Jack.

Sorting It Out

macaroni noodles
pipe cleaners
paint
markers

Sorting

Have the students paint one side of the noodles and mark the other with a homework subject like writing, reading, math, s.s. (social studies) science. String the noodles on the pipe cleaners and twist into a bracelet. When homework is assigned, they turn the written part of the noodle faceup as a homework reminder at the end of the day.

Sorting It Out

"Where is your math homework?" Mrs. Higgins asked, tapping her foot.

David lowered his head. "Oops."

"You've forgotten another homework assignment?" Mrs. Higgins sounded harsh.

"I remembered to do my writing and reading homework," David offered with a weak smile.

Mrs. Higgins wasn't buying it. "Please remember to write down all your homework. You'll have to do the assignment while everyone else is at recess."

David shuffled back to his desk. He hated missing recess. Could he help it if Mrs. Higgins gave too much homework? He couldn't remember all of it.

As homework was assigned during the day, David wrote it in his planner. After all, that's what the planner was for. But forgetful David ran home after school, and forgot his planner!

"Let me think," he said, rummaging through his backpack. "I have to write some sentences about Abraham Lincoln. Do problems 8-15 on page 66 in math. Uh . . . what else?" He shook his head hoping the rest would pop into his brain. Nothing. "That must have been all."

It wasn't all. David faced Mrs. Higgins again the next day.

"David, would you tie a string around your finger or something? You've got to start remembering all your homework."

David had heard of people tying a string around their fingers to help them remember something, but he was sure if he tried it, he'd forget what the string was for. Ugh! There had to be a way to remember all that homework.

He thought about writing it down and pinning it to his shirt. Nope. That would just cause the other kids to poke fun at him. He could write the assignments down on his hand. Uh-huh. They'd just get smeared when he washed his hands for lunch. David was about to give up when he was struck with a brilliant idea.

That morning during art, David took some macaroni noodles and painted one side of each a different color. On the other side he wrote *Math, Writing, Reading, S.S.* (for social studies) and *Science*. When the noodles dried, he sorted them by their daily schedule, and strung them on a pipe cleaner. Then he twisted them onto his wrist. When Mrs. Higgins gave a homework assignment in math, David flipped the math noodle over. When she gave a reading assignment, he flipped the reading noodle over. And even though the bracelet wasn't too tight, it was snug enough to keep the noodles from flipping back. When David got home that afternoon, he checked his wrist. He had three homework assignments to do.

Mrs. Higgins praised David the next day for turning in all his work. "How on Earth did you manage to remember?"

He showed her his bracelet.

Mrs. Higgins's eyes lit up. "What a wonderful idea!"

And that afternoon, she passed out noodles, paint and pipe cleaners to all the students in her class. They all made their own homework helper.

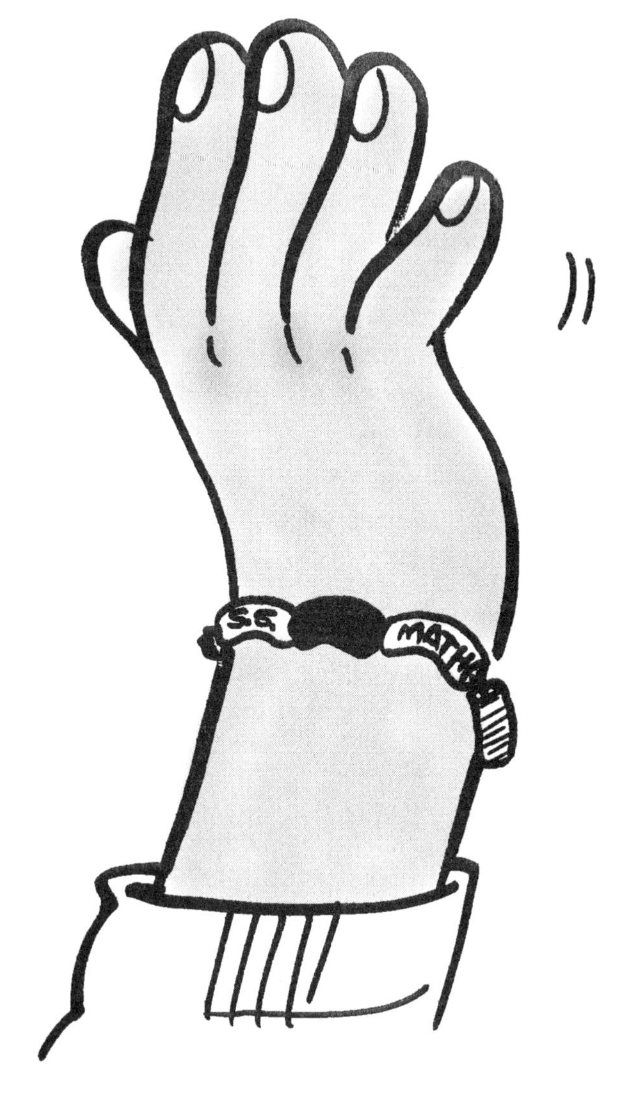

I Want More!

A Calculator Story

calculator with a large display screen

Subtraction

Subtract numbers on the calculator as instructed, and show the difference to the students. At the end of the story, turn calculator upside down to show the word *less*.

I Want More!

A Calculator Story

Julius had the biggest marble collection of anyone in his school . . . maybe even the world! No one, but no one, could match his bags of marbles.

"How many do you have now?" his friend, Scott asked.

"I don't know," Julius said. "I guess I should count them." So Julius did. He counted...and counted...and counted. It seemed to take forever. Finally he counted the last one. Julius had 8,268 marbles.

"Cool!" Scott said. "I only have 36."

While checking on the internet, Julius read that a man in England had collected over 40,000 marbles. "No fair!" he said, kicking the floor. "I'm supposed to have the most marbles. I want more!"

So Julius and Scott set out to find more marbles for Julius's collection. Heaving the bulky bags of marbles along with them, they entered a specialty shop that sold rare marbles.

"Look at these!" Scott said, pressing his face against the display case. "They have some awesome marbles."

Julius stared into the case. He wasn't eyeing all the marbles. Just one in particular. "How much for this marble?" he asked the clerk.

The clerk looked down where Julius was pointing and laughed. "More than you can afford, kid."

"But how much?" Julius urged.

The clerk curled his lip. "That's a vintage 1920 agate."

"Vintage?" Scott asked.

"That means 'old,'" Julius whispered to him.
The clerk continued. "That marble costs $20.00."

"Twenty dollars!" both boys yelled at the same time.

The clerk shrugged. "It's an antique."

"That's okay," Scott said to Julius. "There are plenty of other places that sell marbles you can afford."

But Julius couldn't take his eyes off the marble. "How much do you pay to buy marbles?" he asked.

The clerk shrugged again. "What have you got?"

Julius plopped his bags of marbles on the counter and the clerk looked them over.

"Quite a collection you've got here," the clerk said. "I'll let you have that special agate for one of your bagsful."

Scott's mouth dropped. He knew that Julius had divided his marbles into bags of 1000 with the extra 268 in a smaller bag. "But that's 1000 marbles!" Scott said, "Don't do it, Julius."

Too late. Julius was already handing over one of the full bags.

When they left the store, Scott asked Julius, "How many marbles do you have now?"

Julius took out his calculator. *(Subtract the following numbers on calculator as you read and show the difference to the class.)*

"8,268 - 1000 = 7,268 + 1 = 7,269. Wow, I really have some catching up to do now. I want more."

So they set off again in hopes of buying more marbles. They headed straight to the toy store. "I can afford the marbles in here," Julius said.

But no sooner had he said those words when he tripped on a misplaced skateboard, and went whoops! into the air. When Julius hit the ground, one of his bags of marbles hit the floor and popped open. Marbles scattered everywhere. And not just a few, but the entire contents of the bag!

Scott and Julius scrambled to pick them up, but were only able to retrieve 14 of them. The rest rolled under a door.

"Double fudge!" Julius shouted, kicking the floor. When he opened the door to go in, several ladies inside screamed. That's when Julius noticed the sign on the door. LADIES. The marbles had rolled into the ladies restroom.

"How dare you!" the manager said, rushing over. "Get out now!"

So they went back outside, and took out the calculator. *(Subtract the following on the calculator and show the difference to the class.)*

"7,269 - 1000 = 6,269 + 14 = 6,283."

"I don't think you're ever going to collect 40,000 marbles," Scott said.

"Sure I will," Julius argued. "I want more."

He wasn't sure where to look next, but he led Scott down the street, searching for more marbles. As they passed by the florist shop Scott said, "Look!" In the window was a clear glass vase filled with solid colored marbles. The marbles supported the stem of a huge white lily.

"Let's go in," Julius said.

He was in luck. The florist shop had bags and bins full of those marbles in every color of the rainbow. Julius even thought that there could very well be 40,000 marbles just in that flower shop. "How much are these marbles?" he asked.

A lady arranging a bouquet of roses looked over. "$1.00 per 100 marbles."

Scott nudged Julius. "How much money do you have?"

Julius dug into his pocket and pulled out $2.33.

"That's not much," Scott said.

"But I have to replace my missing marbles. Even if it's just a few at a time. I want more."

Just as Julius began counting out the new marbles, a clumsy dog came racing into the flower shop, snatched one of Julius's marble bags, and dashed back out the door. "Stop!" the boys shouted, chasing the dog.

They finally caught up with him at the park, where he'd torn open the bag, scattering marbles everywhere. When the boys reached down to pick them up, the dog took the bag with the remaining marbles and ran away.

"Fiddle sticks!" Julius cried. "How many did you pick up?

Scott counted his marbles. Between the two of them, they'd only recovered 254 marbles.

"I want more!" Julius shouted.

Scott shook his head. "Forget it."

Julius and Scott went back to his house with the marbles he had left. Julius pulled out his calculator. *(Use the calculator and show the class.)*

"6,283 - 1000 = 5,283 + 254 = 5537."

"You wanted more," Scott said. "But what did you end up with?"

Julius hung his head and turned the calculator over. *(Turn calculator upside down, 5537 looks like the word **less**.)* "Less."

Name ________________________________

The Canine Code

Abby got a puppy for her birthday. He was white with splotchy golden yellow spots. But what could she name him? She thought and thought and thought . . . the perfect name would come to mind. The dog was well behaved except in one area—housebreaking! Abby tried everything. She managed to teach him a few tricks. She taught him to sit, to fetch and to roll over. But no matter how hard she tried, she just couldn't get that doggie to potty outside. "I give up!" she sighed. "You'll never be housebroken, and you'll never have a name." Then she looked at the spots on his fur, and the spots he made on the floor, and smiled. "I have the perfect name for you!"

Can you guess the name? Do the math problems below, then write the answers on the lines. Use the alphabet code to learn the name of Abby's dog.

1. 5 + 11 =
2. 15 + 6 =
3. 10 - 6 =
4. 8 - 4 =
5. 8 + 4 =
6. 11 - 6 =
7. 12 + 7 =

Answers: _____ _____ _____ _____ _____ _____ _____

The dog's name is _____ _____ _____ _____ _____ _____ _____.

Alphabet Code

A = 1	E = 5	I = 9	M = 13	Q = 17	U = 21	Y = 25
B = 2	F = 6	J = 10	N = 14	R = 18	V = 22	Z = 26
C = 3	G = 7	K = 11	O = 15	S = 19	W = 23	
D = 4	H = 8	L = 12	P = 16	T = 20	X = 24	

The Magic Box

chalkboard
chalk

Multiples of 10

When instructed, draw ovals on the chalkboard to represent the potatoes, dollar signs for the dollar bills and triangles for the skunks. Space and group them by sets of 10.

The Magic Box

There once was a man and woman who were so poor they had practically nothing but an old shack and a garden of potatoes. Every day the man would go out and dig up a couple of potatoes for his wife to cook.

One day, after digging up three potatoes, his small spade hit something hard under the ground. The man dug and dug, then realized he'd found a box buried under his potato garden. The man tried to lift it out, but it was too heavy.

"Quick!" he called, running inside to tell his wife. "Come look at what I found."

The woman hurried out beside her husband to see the box. "Well, open it," she said, anxiously.

The man handed her the potatoes, then pried the lid off the box. They both peered inside, only to find it . . . empty.

"Oh well," the man said. "I should have known it would turn out to be nothing at all."

But before he could close the lid, the woman accidentally dropped one of the potatoes she was holding, and it fell into the box. *(Draw an oval on the chalkboard.)* Miraculously that one potato turned into 10 potatoes! *(Draw nine more ovals making a group of 10.)*

"Wow! Did you see that?" the woman asked.

The man was so surprised, he just nodded his head.

The woman dropped in another potato. Ten more potatoes appeared. *(Repeat, drawing 10 more ovals.)* "It's wonderful!" she said. "The box multiplies everything 10 times! One potato makes 10. Two potatoes make 20." *(Repeat, drawing ovals.)*

"So what happens if there are three potatoes in the box?" the man asked. He took the third potato from the woman and dropped it in. *(Repeat.)* Ten more potatoes appeared. "Three potatoes make 30 potatoes."

"This is fantastic," the woman said. "But I'm hungry. Let's eat some of these potatoes now." She reached in and picked one up. *(Erase one oval.)* Pop! Another potato appeared in it's place. *(Draw another oval in its place.)* She smiled. "We'll never run out of potatoes."

During that week the woman made every potato recipe she could think of. Mashed potatoes, baked potatoes, scalloped potatoes, French fries, hash browns and potato soup. Finally she said, "I'm sick of potatoes. Why don't we take a bag of them to sell at the market. Then we can buy other things to eat."

That's just what they did. Every day they scooped up more potatoes and sold them at the market. They bought lots of food and still had plenty of money left over.

One afternoon the man and woman came in from the market, and before going inside, the woman reached down into the magic box to pick up some potatoes. But when she did, she accidentally dropped a dollar into the box. *(Repeat instructions with dollar signs)* Poof! Ten dollars magically appeared. "Look!" she cried out to the man.
They dropped in another dollar. *(Repeat.)* That one turned into 10 also. "See?" the woman said. "It's multiples of 10 again. *(Repeat.)* One dollar turns into 10 dollars. Two dollars turn into 20 dollars."

Luckily, they had been saving some of the money made at the market. *(Repeat instructions for every dollar dropped in.)* They dropped in a third dollar—poof! Thirty dollars. Four dollars equaled 40. Five equaled 50. Every dollar they dropped in the magic box multiplied itself by 10. No need to sell potatoes at the market now. They only had to scoop money from the box if they needed to buy something. And every dollar they took, magically replaced itself.

They continued to drop money into the box as well, and soon they were extravagantly rich. They built a new house where their old shack had been, and filled it with fancy furniture and lots of servants. Everything was perfect.

Until . . .

One day when the man got careless and didn't replace the lid properly on the magic box. That night a skunk was wandering by and fell inside it. *(Repeat instructions using triangles.)* Pop! Suddenly there were 10 skunks! They fought and clawed to get out, spraying the air with the worst kind of odor. As one crawled out, another appeared. Other skunks in the area sniffed and came over. They would fall or jump into the box. Pop! Ten more skunks! Pop! Twenty skunks! Pop! Thirty skunks! Pop! Forty skunks! Pop! Fifty skunks!

The woman jerked awake and shook her husband. "Ew! Something stinks."

"It smells like a skunk," the man said.

"It smells like a million skunks," the woman added.

They put some clothespins on their noses and ran out to the magic box. "Do something!" the woman shouted to her husband.

He tried pulling them out with a net, but more skunks appeared. He tried dumping some baking soda on them, but that only hides small odors, not skunk odor. He even splashed them with an entire bottle of his wife's expensive French perfume. Nothing worked.

"I can't live here with this disgusting smell," the woman said.

The man nodded in agreement. The skunks' smell had permeated the entire house, the furniture and the couple's closet of clothes. They couldn't sell their home because no one wanted to live in a stinky house. So the man and woman had to leave everything behind. They did get another shack to live in. And in the garden, they only had potatoes.

Mathematical Me

Tape measures

Measurement and addition

Have the students use the tape measures to measure the distance from their knees to the bottom of their foot. Double the amount. Then measure the distance from hip to foot, proving that the measurement is the same as the amounts added. Repeat with knuckles and fingertips to get length of fingers.

Mathematical Me

My teacher, Ms. Willow, is always asking us strange questions. Questions like—"If you were a baked potato, what toppings would you have?" My toppings would be chili and cheese. But no chives. I hate green things on my food.

Mrs. Willow said that chili and cheese make for a zesty potato, and that certainly fit me. Some kids were just plain old potatoes with butter. My friend Mike said he was French fries. Ms. Willow asked him what size—small, medium or large, and why would he be tastier than any other fries? She's a tough teacher!

Today she asked us another silly question. "If the human body were a school subject, which subject would it be?"

Naturally, most kids said science since the human body has veins and blood and other science stuff. Lisa said that the body could be social studies. That answer really intrigued Ms. Willow. I could tell because she placed her pointer finger to her chin as she listened to Lisa's answer.

Lisa said the body was like social studies because people are like their environments. Wherever they live determines how they look, how they talk and what foods they eat.

Ms. Willow said that was such an excellent answer. But I raised my hand because I thought my answer was 10 times better than Lisa's.

"Okay, Paul, what do you think?" Ms. Willow asked.

I didn't hesitate a minute. "Ms. Willow, I think the body would be math."

Ms. Willow put her finger on her chin again. "Please share with us why you think that?"

"It's simple," I tell the class. "If I measure the distance from my knuckle to the joint in the middle of my finger, I can add that same amount to it and that will tell me how long my finger is."

Ms. Willow's eyes lit up like I was one smart dude.

"And that's not all," I said. "If I measure the distance between my knee and the bottom of my foot, and add the same amount to it, that's the length of my leg."

"I don't believe it," Lisa said. I think she was jealous because she thought her answer was the best.

"I'll prove it." I took Ms. Willow's tape measure and had Lisa come to the front of the class. So they wouldn't think I was cheating, I had Ms. Willow measure Lisa's leg from her knee to the bottom of her foot. It was 14 inches. 14 + 14 = 28. Ms. Willow measured Lisa's leg from her hip to her foot. Guess how many inches? Yep, 28. "It works every time," I said. Ms. Willow measured some of the other kids to show it was true.

"So the human body is a walking math problem," I said.

Ms. Willow agreed. And so did everyone else in class as they measured their fingers, arms and legs. *(Have the students try it with the tape measures.)*

Name ________________________________

Fancy Math Words

Draw a line from the word to its meaning.

1. angle
2. chart
3. cube
4. denominator
5. graph
6. googol
7. numerator
8. parallel
9. polygon
10. spiral

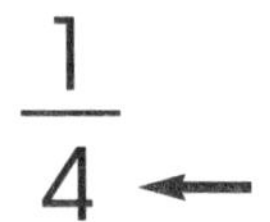

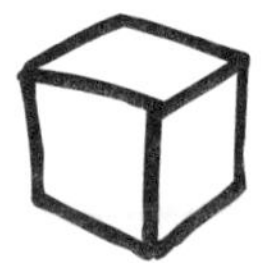

1000000000000000000000000000000000000000000000000
000000000000000000000000000000000000000000000000000

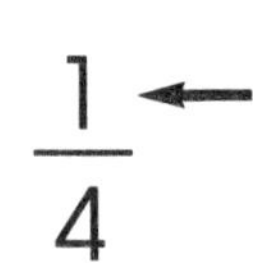

Birthday Party

Invitations	Decorations	Food	Favors
guest list	cups	cake	toy
invitations	plates	punch	candy
stamps	napkins		gum
	balloons		

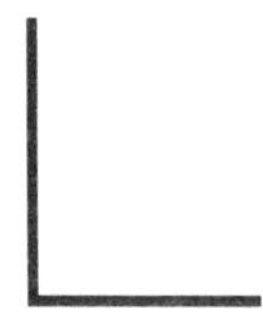

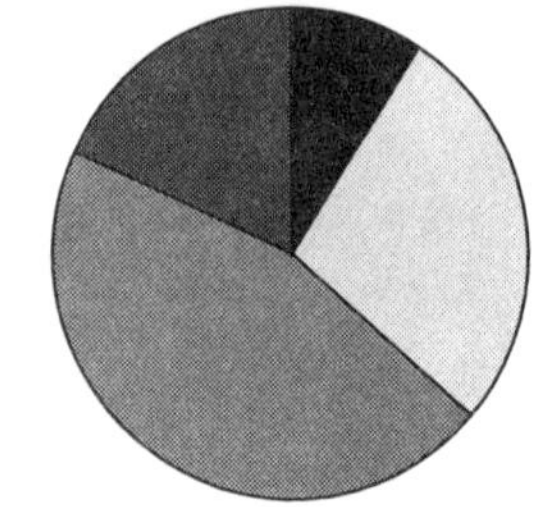

Name ___________________________________

Famous Shapes

Complete the missing name by writing in the shape.

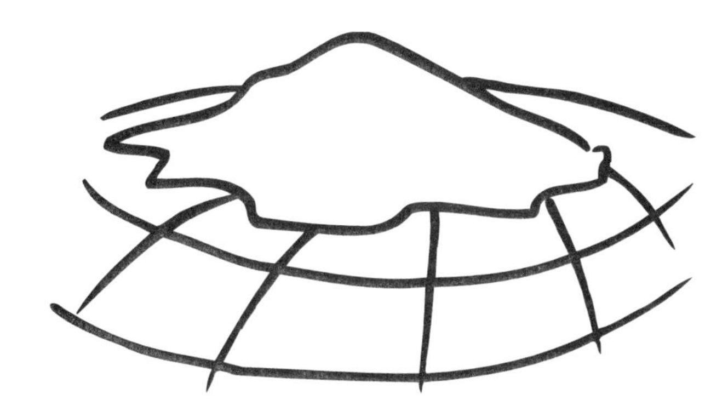

1. Arctic ___________

2. Ice _________

3. Border__________

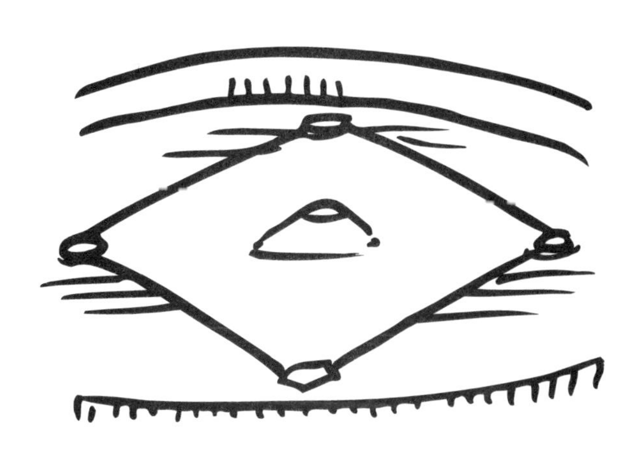

4. Baseball __________

5. Bermuda __________

6. Times ___________

The Cricket's Song

small plastic bottle
with a few beans
closed inside
stopwatch

Timing and addition

Turn the bottle of beans over and over slowly to imitate a cricket sound. At the end of the story use the stopwatch to time 15 seconds. Turn the bottle over and over and have the students count the number of rattles. Have them write it down, then add 40 to that number to find out how crickets can tell us the temperature.

The Cricket's Song

Terry loved visiting his grandpa's house at Circle Lake, but the woods surrounding the lake were so noisy.

"I can't go to sleep," Terry said to Grandpa.

Grandpa smiled. "Well, you are used to your own room at home, and your own bed and pillow."

"I brought my pillow with me," Terry said.

Grandpa tucked Terry in, but didn't bother with the quilt.

"Grandpa! Aren't you going to cover me with more than this? I'll get cold in the night."

Grandpa leaned an ear toward the window screen and listened. He looked down at his watch, and nodded his head in a rhythm like he was silently counting. *(Turn the bottle over slowly several times to reproduce a cricket sound.)* After a series of nods, he looked up at Terry and said, "You won't need the quilt tonight. It's really warm out. And from the sound of the crickets, I doubt the temperature will drop much."

Grandpa leaned over and gave Terry a kiss on the forehead, turned out the light and closed the door behind him.

Terry lay alone, listening to the crickets. Their loud chirps sounded more like rattling than bug calls. Then the frogs joined in, along with a nearby owl. Terry put his pillow over his head and eventually went to sleep.

When Terry woke up, the sun was sending a warm golden glow through the window. The night noises had been replaced with a symphony of birds.

Terry spent another exhausting day with Grandpa. They hiked, fished and swam in the clear lake. When he went to bed that night, he didn't think anything could keep him awake. But while Grandpa tucked him in, Terry heard a lone cricket rubbing his legs together in the loudest chirp imaginable. It sounded like an awful ringing in his ears. *(Turn the bottle over a few times to reproduce the cricket sound.)* "Grandpa, would you please kill that bug?"

Grandpa shook his head. "No way. A cricket in the house is good luck. But I'll take him out if that will help."

Grandpa scooped up the cricket in his hands and headed for the door. "And besides," he said while walking out, "if I killed the crickets, I wouldn't know what the temperature is outside." He clicked off the light with his elbow as he cuddled the cricket in his hands.

The temperature? Terry wondered. The day had caught up with him, and he was just too tired to try and figure it out.

The next night proved as noisy as the others. "Grandpa, how can you stand it? There must be a million crickets out there."

"Yep," Grandpa said, "which reminds me." Grandpa held up a small bag. "I bought this while we were in town today." He pulled out a small wall thermometer. "Most folks read the numbers on those things to tell how hot it is. But I have a better trick. Now stay real quiet."

Terry did. He watched while Grandpa nodded his head in time with the crickets. *(Turn the bottle over slowly to represent the cricket's chirp.)* Then Grandpa stopped nodding, and looked at him.
"I bet it's about 75 degrees outside . . . give or take one. Right?"

Terry looked at the thermometer. "Wow! You're right. How did you do that?"

Grandpa's mouth slanted into a sly grin. "It's all in the math."

Terry sat up anxiously. "Tell me."

"Listen to the crickets," Grandpa said, taking off his watch. "Use this to time it. Count the number of chirps for 15 seconds."

Terry took the watch and did what Grandpa said. *(Turn the bottle over for the cricket sound.)* "Thirty-five," he said. "I counted 35 chirps."

"Now add 40 to that number," Grandpa instructed.

Terry did the math in his head. 35 + 40 = 75. "Wow! Grandpa, it's 75 degrees outside!"

"Give or take one," Grandpa said with a wink.

"Does this work every time?" Terry asked.

Grandpa nodded. "Yep. And remember, crickets chirp slower in cooler weather and faster in warm weather. But if you do the math, you'll always have a good idea of the temperature."

Terry didn't mind the crickets anymore. He spent the next few nights at Grandpa's house timing the chirps and adding 40. Then he'd tell Grandpa the temperature.

And when he went back home at the end of the week, Terry felt quite clever. He told all his friends about his cricket math trick.

Trading for Eggs

A Modern Aesop Fable

copies of reproducible page 62

Translation and puzzle solving

After reading the story, have the children decipher the moral of the story by translating the coded numbers to the alphabet letters on the phone buttons.

Trading for Eggs

A Modern Aesop Fable

Marshall lived in a small apartment in the middle of town. That's why it was a fun adventure for him when he rode the school bus home with his friend, Luke. Luke lived on a real working farm.

One day while Marshall helped Luke gather eggs from the hen house, Luke said, "Here. Take this egg home with you. Keep it safe and warm, and in a few days you'll have a baby chick."

"But we're not allowed to have pets in our apartment," Marshall said.

Luke grinned. "That's okay. You'll keep the chick in a box. No one will know."

So Marshall took the egg home, and set it down on a pillow inside a cardboard box. He shined his desk lamp on it to keep it warm. As Marshall sat looking at the egg, he began to think. One chick could be fun, but if I had two chicks they would lay plenty of eggs, and I'd get more chicks, and I could sell them and make lots of money.

Marshall loved his clever idea. He reached for the phone and called Luke. "Hey . . ." he began, "do you think you could give me another egg so I'll have two chicks to keep each other company?"

There was silence on the other end of the line, then Luke said, "Hmmm . . . I don't know, Marshall. My dad might get mad if I give away any more eggs."

"I'll trade you for another one."

More silence.

"What would you trade?" Luke asked.

Marshall paused to think. "How about some baseball cards?"

"Okay," Luke answered. "But I get to pick which ones."

Marshall didn't mind. He'd just buy more baseball cards with the money he made off his chickens.

The next day, Marshall rode the bus with Luke to his farm and traded three baseball cards for another egg. When he got home, he placed it on the pillow with the other egg.

Marshall lay back on his bed with a huge grin. I'm going to have so many chickens to sell, he thought. But he sat up quickly when another thought occurred to him. What if these eggs are both roosters? Roosters don't lay eggs!

Marshall grabbed the phone. "Luke, I have a problem. I need another egg."

"No way!" Luke said. "My dad will kill me."

"Please," Marshall pleaded. "Just one more."

"Sorry," Luke said. "I can't."

Marshall sighed. "I'll trade you for it."

"But I have all the baseball cards I want," Luke said.

Marshall thought about it. What could he offer Luke to get another egg? "How about I trade you my baseball glove. It's 10 times better than yours."

"You got it!" Luke said, sounding like he'd won the lottery.

The next day, Marshall took his baseball glove to school, and then traded it to Luke that afternoon at the farm. Marshall brought home his egg and set it down on the pillow. He looked at the three eggs. Surely one of the three had to be a hen. But what was the guarantee? He called Luke on the phone. "I need more eggs."

"Why?" Luke asked. "Are you starting a restaurant?"

"No," Marshall said. "I just need several more."

"I think you just want to get me into trouble," Luke said. "I wish I'd never given you an egg in the first place."

"Come on," Marshall argued. "I'd really like to have about a dozen eggs here."

Luke laughed. "Yeah, and my dad will ground me forever!"

"I'll give you anything," Marshall said.

Luke was very quiet on the other end. "Anything?" he whispered.

"Anything."

The next day Marshall traded his $100.00 fishing rod for nine eggs. Nine plus the three he had equaled twelve eggs. One dozen. He didn't mind because he knew he could sell those chickens and make oodles of money to replace everything he traded.

When Marshall placed the eggs on the pillow, two rolled against each other and cracked. "Oh no!" he said, trying to clean it up. The gunky yokes stuck to his fingers. The pillow was a yellow mess. He reached for the pillow to lift it out, eggs and all, but the remaining eggs slipped off, and landed on the bottom of the box with a nasty splat! All 12 eggs were as cracked and broken as Humpty Dumpty, and Marshall knew no one could put them back together again.

Marshall reached for the phone. "May I speak to Luke, please?"

"Sorry," his dad said. "Luke went fishing with his new fishing rod. He'll probably be out for a while because he plans to sell the fish to make some extra money."

Why didn't I think of that! Marshall thought. "Thanks," he told Luke's dad.

Marshall stared at the buttons on the phone, wanting to cry. There is a moral to his story, and he'd definitely learned it the hard way.

Do you know the moral?

Using the keypad on the picture of the phone, decipher the moral to Marshall's story.

Name ___________________________

Trading for Eggs

Write in the letter that is shown on the dial buttons of the phone by matching it to the numbers listed. The first number shows which button to use, and the second number shows the placement of the letter. For example: 2-3 means "C."

__ __ __ __   __ __ __ __ __   __ __ __ __
3-1 6-3 6-2 8-1   2-3 6-3 8-2 6-2 8-1   9-3 6-3 8-2 7-2

__ __ __ __ __ __ __ __   __ __ __ __ __ __
2-3 4-2 4-3 2-3 5-2 3-2 6-2 7-3   2-2 3-2 3-3 6-3 7-2 3-2

__ __ __ __   __ __ __ __ __.
8-1 4-2 3-2 9-3   4-2 2-1 8-1 2-3 4-2

Name ________________________________

Ridiculous Arithmetic Riddles

1. If you're trapped in a box with just a stick, how can you get out? ____________________

2. How many sides does a box have? ____________________

3. A woman has 5 children, and half of them are boys. How is that possible?

4. Why was the math book crying? ____________________

5. Why is the longest human nose only 11 inches? ____________________

6. A mother and father have 4 sons, and each son has a sister. How many people are there in the family? ____________________

7. What do you get when you add two apples and three apples? ____________________

8. Why did 6 run away scared? ____________________

9. What did the teacher say to the boy who didn't want to take his math test?

10. Where do math teachers eat their lunch? ____________________

Answer Key

St. Ives, page 24

One

The Canine Code, page 44

Answers to math problems:

1. 16 2. 21 3. 4 4. 4 5. 12
6. 5 7. 19

The dog's name is PUDDLES.

Fancy Math Words, page 52

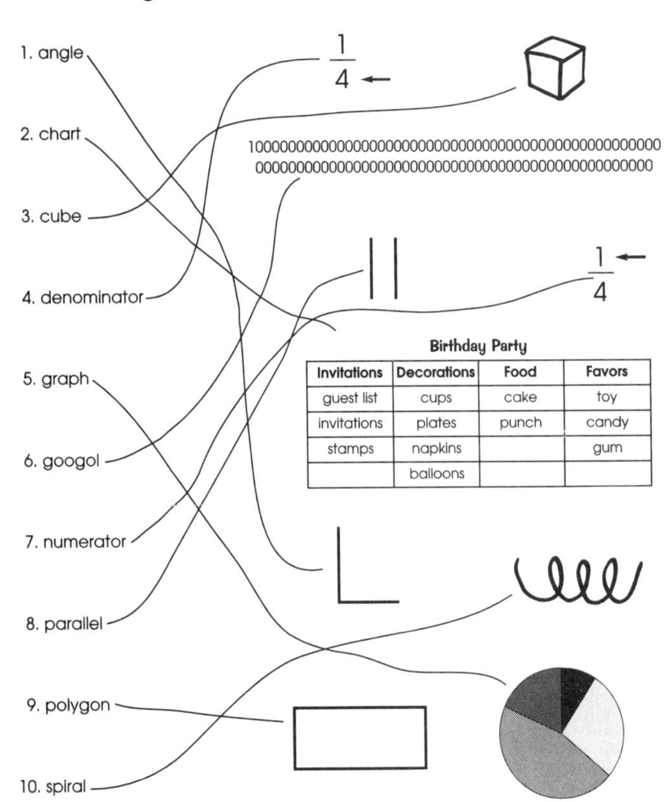

Famous Shapes, page 53

1. Arctic Circle
2. Ice cube
3. Border line
4. Baseball diamond
5. Bermuda Triangle
6. Times Square

Trading for Eggs, page 62

Don't count your chickens before they hatch.

Ridiculous Arithmetic Riddles, page 63

1. Break the stick in half; two halves make a whole, so crawl out of the whole.
2. Two—inside and outside.
3. The other half is also boys.
4. He had so many problems.
5. If it had another inch, it would be a foot.
6. Seven. There is only one sister.
7. A math problem
8. Because 7 ate 9.
9. Go figure!
10. On a multiplication table